Reading Essentials®
in Science

ENERGY WORKS!

Light

JENNY KARPELENIA

PERFECTION LEARNING®

Editorial Director:	Susan C. Thies
Editor:	Mary L. Bush
Design Director:	Randy Messer
Book Design:	Mark Hagenberg, Michelle Glass, Lori Gould
Cover Design:	Michael A. Aspengren

A special thanks to the following for their scientific review of the book:

Judy Beck, Ph.D; Associate Professor, Science Education; University of South Carolina-Spartanburg

Jeffrey Bush; Field Engineer; Vessco, Inc.

Image Credits:
©Bettmann/CORBIS: p. 5 (right); ©John Farmar;Cordaiy Photo Library Ltd/CORBIS: p. 21;
©Mark A. Johnson/CORBIS: p. 25

©Royalty-Free/CORBIS: p. 20 (bottom and middle); ArtToday (arttoday.com): pp. 4, 5 (left), 29 (middle), 35 (top), all art on sidebars; Photos.com: all cover art, pp. 6, 7, 9, 10, 11 (bottom), 12, 14, 15, 18, 20 (top), 22, 23 (top), 24, 26, 29 (top), 30 (bottom-left background), 31, 32, 33, 34, 35 (bottom); Life ART©2003 Lippincott, Williams and Wilkins: pp. 27, 28; Corel: back cover, pp. 2–3; Lori Gould: pp. 8, 19; Perfection Learning Corporation: pp. 11 (top), 13, 16, 17, 23 (bottom), 30 (bottom diagram), 36

For information, contact
Perfection Learning® Corporation
1000 North Second Avenue, P.O. Box 500
Logan, Iowa 51546-0500.
Phone: 1-800-831-4190
Fax: 1-800-543-2745
perfectionlearning.com

3 4 5 6 7 8 PP 10 09 08 07 06 05

Paperback ISBN 0-7891-5864-7
Reinforced Library Binding 0-7569-4450-3

Contents

Introduction to Energy

ENERGY—WHAT IS IT?

Was there ever a time when you felt so tired that you could not even shoot some hoops or play your favorite video game? Maybe you were feeling sick or very hungry. You probably felt as if you had no energy.

Now think of a time when you had lots of energy. You felt as if you could run, talk, ride your bike, or play games forever. Perhaps your parents or teachers even told you that you had *too much* energy.

So what is energy? What forms of energy are there?

Energy is the ability to get things done or to do work of some sort. Anything that accomplishes something is using some form of energy. When you hear the word *work*, do you think of chores around the house? *Work* actually means getting *anything* done. A football sailing through the air is using and giving off energy. A ringing doorbell is using and giving off energy. A shining lightbulb is using and giving off energy.

These examples also show some of the different forms of energy. A thrown football is an example of motion energy. A ringing doorbell is using electrical energy and giving off sound energy. A lightbulb uses electrical energy and gives off light and heat energy. Motion, electricity, sound, light, and heat are all forms of energy. These forms of energy affect our lives every day.

Carnival rides use electrical energy to create motion energy.

Albert Einstein

IT'S A LAW

Scientists perform experiments to test their **theories**, or ideas, about the world. Experiments that produce the same results over and over become scientific laws. One of these laws says that energy cannot be created or destroyed, but it can change from one form to another. This means that all around us, every day, energy is being changed from one form to another. The amount of energy in the universe stays the same, but it is constantly taking different forms.

The famous scientist Albert Einstein was a great thinker. He thought of new ideas that other people had not even imagined. He developed the equation $E = mc^2$. The E stands for "energy." The m stands for "mass" (the amount of "stuff" in an object). His idea shows that energy and mass can change back and forth. So energy can be changed into stuff, and stuff can be changed into energy.

ENERGY WORKS!

Energy is very important. It allows many types of work to be done. People have energy. Plants have energy. The Sun gives off energy. Machines use energy. Read on to find out more about energy, its forms, how it works, and how it is used.

CHAPTER 1
Lighting Up
Your Life

Crowds of people cheer from the stands. A player dressed in a sparkling blue and white uniform crosses the plate. The home team has just scored another run.

Music blares over the loudspeakers. The scoreboard lights up. Bright red numbers show the new score. Streams of colorful messages streak across the scoreboard.

The weather is perfect for a baseball game. It is warm and sunny. Many of the fans around you are wearing sunglasses. Sunlight glints off shiny aluminum soda cans and wrappers.

The next batter is up. You can barely see the white of the ball as it whizzes by. The catcher's brown glove cradles the strike. Fans from the visiting team roar. They wave red banners in support of their team. The game is on!

Light surrounds us. A day at the ballpark is full of light. Everything you see is because of light energy. The colorful uniforms, the brightly lit scoreboard, and the ball as it flies through the air are all visible because of light.

This important form of energy lights up your life and the world around you.

CHAPTER 2
Waves or Particles?

People have argued about light for many years. Some scientists think that light travels as a **wave**. Other scientists think that light travels in a stream of **particles** called *photons*. Today, many scientists think light acts as both a wave and photons.

WAVES

A transverse wave has high points and low points, like a water wave. Light travels as a transverse wave. Radio waves, **infrared** energy, visible light, **ultraviolet** light, microwaves, and X rays are some other types of energy that travel as transverse waves. These forms of energy are called *electromagnetic radiation*. Visible light is the only type of electromagnetic radiation we can see with just our eyes. We need machines to help us see the other types of energy.

Scientists can measure many qualities of light waves. Some qualities are the wavelength (distance between waves), frequency (number of waves per second), and speed.

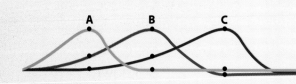

Transverse Wave

PHOTONS

Objects appear to give off light when their **atoms** are energized. Atoms are energized when they are heated, electricity is passed through them, or they undergo a **chemical** reaction. When these energized atoms return to normal, they give off light particles, or photons. Photons of different energy levels give off different-colored light.

LIGHT TRAVEL

Light can travel through substances called *mediums*. Mediums are materials that allow other materials to pass through them. Air, glass, and plastic are a few of the mediums that light can move through. Light can even travel through empty space, which is called a *vacuum*. If light could not travel through a vacuum, it would not be able to get from outer space to Earth.

Light travels extremely fast. Scientists have developed ways to measure how fast light travels. The speed of light is approximately 186,000 miles per second.

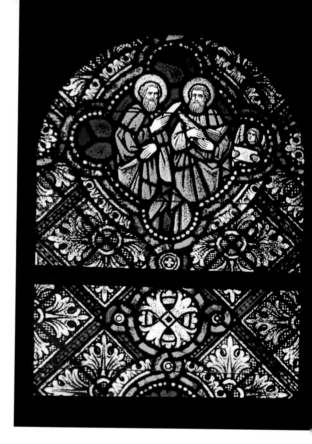

Albert Einstein experimented with light. He often asked the question "What if . . . ?" He wondered what would happen if people and other objects could travel as fast as light or faster. He believed that if people could travel that fast, time would slow down. He also thought space would get squashed. His ideas have not yet been proven right or wrong. So far, nothing is known to travel faster than light.

CHAPTER 3

Bouncing Light

Light is able to bounce off surfaces. This is called *reflection*. You are able to see things because light hits objects and some of that light bounces, or reflects, back to your eyes.

Light bounces off objects in straight lines. These lines hit and reflect off a surface at the same angle. The direction the light travels depends on the surface.

When you look into the smooth surface of a mirror or glass, you see a clear reflection. This is because the light strikes the surface and bounces right back to your eyes. Think about staring into the smooth surface of a lake. The light bounces straight back at you, so you see a clear reflection. This is a **regular reflection**.

However, if the surface reflecting the light is not smooth, the light bounces, or scatters, in all different directions. This makes the reflection fuzzy or unclear. Imagine staring into the surface of a lake after someone has jumped in and made waves. The reflection you see is fuzzy. This is an **irregular reflection**.

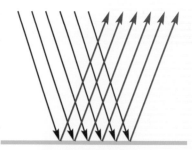

**Regular Reflection
(smooth surface)**

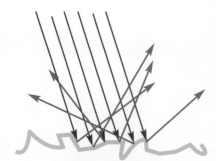

**Irregular Reflection
(rough surface)**

Take a piece of smooth aluminum foil. Look at the shiny side. Can you see your reflection? How does it look— clear or fuzzy?

Now crumple and uncrumple the aluminum foil. This makes the surface uneven and bumpy. Look at your reflection again. Is it clear or fuzzy?

CHAPTER 4
Bending Light

REFRACTION

When light travels through materials, it will either speed up or slow down depending on the medium. A denser material slows down light. A less dense material speeds up light.

As light travels from one material to another, it changes speed and bends. This bending of light at an angle as it passes through materials is called *refraction*.

A Definition for *Dense*

Dense is a word used to describe how tightly or loosely packed a material's atoms are. Some materials, such as glass or wood, have tightly packed particles and are considered dense. Water and air have loosely packed particles that flow freely, so they are less dense than other objects.

For example, when light passes from water to air, it speeds up and bends. Air is less dense than water, so the light can speed up when it hits the air. Some light bends to our eyes and allows us to see objects under the water. But the objects are not really where they seem to be. They are actually in front of where we see them. This is because of the angle that the light is bending.

When light travels from air to water, the light slows down and bends the opposite way. Water is denser than air, so the light must slow down.

Clean out an empty plastic container, and tape a coin to the bottom of the inside. Walk backward slowly until the coin is just out of sight. Now have someone else slowly pour water into the container. Can you see the coin again? Why?

The light bounced (reflected) off the coin and traveled through the water. The light then moved through the air and bent (refracted) back to your eyes. The coin hasn't moved at all. The water just changed the speed of the light and, therefore, its angle to your eyes.

Refraction
The light bends to allow us to see the pencil under the water. But the pencil is not where it appears to be because the light is refracted as it passes through the water.

LENSES

People can make objects from glass or plastic that bend light. These objects are called *lenses*. Lenses are **transparent** materials that allow light to pass through and bend. We can use them to help us see better. Eyeglasses, contact lenses, microscopes, telescopes, and cameras are a few tools that have lenses.

The shape of the lens determines which way it bends light. Light bends toward the thicker part of the lens. Lenses that are thicker in the middle than at the edges are called *convex*. Convex lenses help bend the light rays inward to focus at a point. Magnifying glasses, cameras, overhead projectors, and your eyes have this type of lens.

Lenses that are thinner in the middle than at the edges are called *concave*. Concave lenses help bend the light rays outward. The light rays travel away from one another. Concave lenses are usually used with convex lenses to help clear up the image.

Eyeglasses use convex and concave lenses to solve different problems. They sometimes use a combination of lens shapes to focus light to help a person see. One side of the lens may be curved inward while the other side is curved outward. This combination lens is called a *concave-convex lens*.

CHAPTER 5
Sunlight

The Sun is just one of billions of stars in the universe. Although it is about 93 million miles away, the Sun is the closest star to Earth. That is why the Sun appears much brighter than other stars.

The Sun is mostly made up of a gas called *hydrogen*. This gas is its fuel. The hydrogen that makes up the Sun is under extreme high pressure and temperatures. Under these conditions, the hydrogen atoms fuse, or join, together to form a different gas called *helium*.

Journey to Earth

Sunlight takes about eight minutes to reach Earth from the Sun.

The changing of one type of atom to another is called *atomic energy*. Atomic energy can change into heat and light energy. The warmth and light of the Sun that humans experience on Earth comes from this energy.

The Sun not only provides a natural source of light and heat, but it is also responsible for times of day and the seasons.

DAY AND NIGHT

At any given time, different places on Earth are receiving different amounts of sunlight. This is due to the Earth's shape and its **rotation**.

The Earth is a sphere. It is shaped like a ball, but not a perfect one. The Earth rotates, or spins, on its **axis** once a day. The Sun can only shine on half of the Earth at a time. One half of the Earth is facing toward the Sun and experiencing day. The other half of the Earth is facing away from the Sun and experiencing night.

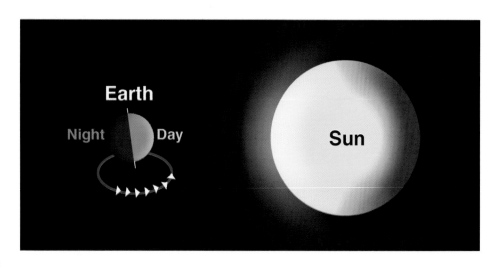

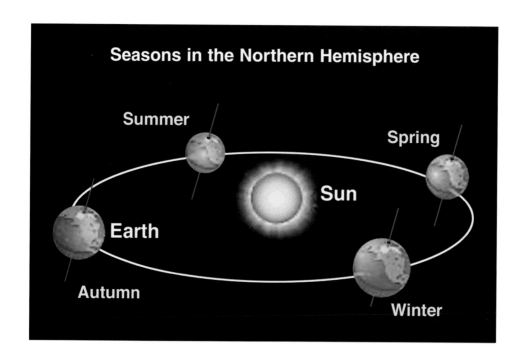

Seasons in the Northern Hemisphere

Summer

Spring

Sun

Earth

Autumn

Winter

SEASONS IN THE SUN

The seasons are due to the tilt of the Earth as it **revolves,** or travels in its orbit, around the Sun. When the Northern **Hemisphere** of the world is tilted toward the sun, the Northern Hemisphere receives more direct sunlight and more heat. As the Earth rotates, the Sun appears to pass overhead. The days grow longer, and the nights are shorter. This season is summer.

When the Northern Hemisphere is tilted away from the Sun, it receives less direct sunlight and less heat. The Sun appears lower in the sky, making the days shorter and the nights longer. This time of the year is winter.

Spring and fall are seasons of fairly equal day and night hours. During these seasons, the Northern Hemisphere receives a **moderate** amount of direct sunlight.

The seasons in the Southern Hemisphere occur at opposite times of the year from the seasons in the Northern Hemisphere.

CHAPTER 6

Light Energy from the Sun

The Sun is the main source of light energy on Earth. All life on the planet depends on this energy source.

THE LIFE CYCLE

Plants use light to help them grow. The light from the Sun changes carbon dioxide and water into sugar and oxygen. Plants use the sugar as fuel to grow. The oxygen is given off as a waste product. This process is called *photosynthesis*.

While the oxygen cannot be used by plants, animals and humans need it to survive. They breathe in the oxygen, which is used to burn the food they eat. Their bodies also use the energy for growing, moving, and communicating. Both animals and humans give off carbon dioxide gas as a waste product. That gas is used by the plants. This life cycle is all dependent on the light energy from the Sun.

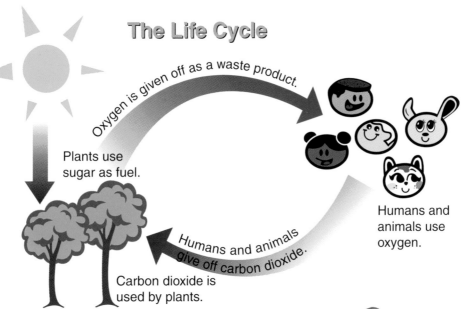

The Life Cycle

Oxygen is given off as a waste product.

Plants use sugar as fuel.

Humans and animals give off carbon dioxide.

Carbon dioxide is used by plants.

Humans and animals use oxygen.

HEALTH ISSUES

Light is necessary for proper health in human beings. Light helps humans use the vitamin D found in milk and food products. Vitamin D helps bones grow.

Sunlight gives skin a healthy glow when absorbed in proper amounts. Too much sun can be harmful to skin and eyes.

Some people feel that sunlight contributes to a healthy emotional well-being. A bright, sunny climate often helps people feel more positive, while a darker environment may cause sadness and depression.

Solar = Sun

The word *solar* comes from the Latin word *solaris*, which means "coming from or relating to the Sun." This important word has similar forms in other languages. *Sol* is the Spanish word for "sun," while *soleil* means "sun" in French.

Solar-powered calculator

Solar cells are organized into arrays to provide electric power to residences and businesses.

Solar power plant in California

SOLAR ENERGY

Light energy from the Sun can also be changed into other forms of energy. For example, light energy can be changed into electricity. Solar cells, also known as photovoltaic cells, change light energy into electricity. The light from the Sun causes electrons to flow in these cells, creating a **current** of electricity.

Solar energy is used in some calculators. A small panel on the machine absorbs light and changes it into electricity to run the calculator.

Larger sizes or numbers of solar cells can change sunlight into much greater amounts of electricity. This electrical energy can then be stored in batteries or sent out across power lines to homes and businesses.

Sunlight can be changed into heat energy. Solar panels on a roof are usually painted black. Black **absorbs** the light from the Sun. The light energy is then changed to heat energy, which warms the home.

Solar panels can be used to heat water for homes. Heat is absorbed by water in pipes running through the solar panels. This creates hot water for the home.

Mirrors can also be used to heat water and produce electricity. Solar furnaces use mirrors to focus light on a water tank. This heats the water to boiling. The boiling water produces steam. The powerful steam turns a **turbine** to generate electricity.

Solar energy is a renewable energy source. That means it can be used over and over again. Scientists are always looking for ways to save energy and not pollute. Solar energy may be used more in the future as the supply of fuels like coal, oil, and natural gas run out.

Solar panels on a solar-powered car

Race for the Sun!

Solar-powered cars are run by special panels that turn solar energy into electricity. Some of these cars use part of the energy to run while saving some in a battery for use when it's cloudy. Others use all of the energy to run and cannot move when the Sun isn't out. Either way, solar-powered cars don't waste other fuels or cause pollution. Today, most of these cars are used for racing or research. Someday, however, everyone may drive solar-powered cars!

CHAPTER 7

Other Sources of Light

Imagine if sunlight were the only source of light on Earth. What would happen after the Sun went down? Luckily, there are other sources of light energy besides the Sun.

FIRELIGHT

Matches, candles, and other lighting tools give off light energy by burning chemicals or other substances that produce a fire or flame. These tools are used as independent sources of light or as a means to create another form of light energy. For example, a match can be a limited source of light as it burns, or it can be used to light a fireplace, which becomes a new light source.

All of these light sources also give off heat energy as they burn.

INCANDESCENT LIGHT

Lightbulbs produce light with the help of electricity. In an incandescent lightbulb, electricity passes through a very thin wire made of a metal called *tungsten*. This creates a lot of heat. The wire glows and gives off light. This is why the bulb is called *incandescent*, which means "glowing."

Only a small percentage of the electrical energy (about 10 percent) that goes into the bulb is turned into light energy. Most of the electrical energy (about 90 percent) that goes into the bulb is wasted as heat.

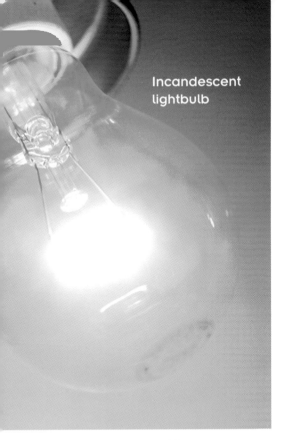

Incandescent lightbulb

bulbs are glass tubes filled with a mixture of argon and mercury gases. The tube is coated with a chemical paint. When electricity passes through the bulb, the gas is energized and reacts with the paint. The coating of chemical paint glows and gives off light.

Fluorescent bulbs are much more efficient than incandescent bulbs. This means that much more of the electricity that goes into the bulb is turned into light, not wasted as heat. Therefore, fluorescent bulbs save money and energy.

FLUORESCENT LIGHT

Another type of lightbulb is a fluorescent bulb. Fluorescent

NEON LIGHT

A third type of lightbulb is called a *neon light*. Neon is a gas.

Fluorescent lightbulb

Edison Invents Incandescent Lightbulb!

Thomas Alva Edison invented the first lightbulb in 1879. Edison had experimented with the various parts of a bulb for years before finally putting them all together to create his incandescent bulb. Edison also invented the telegraph and phonograph.

It can be placed in glass tubes of many shapes. When electricity passes through these tubes, the gas glows and gives off a reddish orange light.

LIVING LIGHT

Some living things have the ability to produce their own light. This is called *bioluminescence*. A chemical reaction creates a flash or glow of light in the bodies of these creatures.

Some fish that live in the deep, dark ocean give off their own light. Fireflies flash their lights at one another to communicate. They also use the light to signal that they are ready to **mate**. Other animals use their light to scare off predators or attract prey.

Lights of Other Colors

Other gases can be used to produce various colors of light. In a clear glass tube, argon gives off a light purple glow. Krypton creates a white light. Xenon glows blue. If the glass tube isn't clear or is coated inside, other colors or **shades** of light are produced.

CHAPTER 8

Light to Our Eyes

Our eyes are complicated sense organs. They are made up of many parts that work together. Our eyes gather light, focus light, notice colors, and send information to the brain.

THE EYE EXPERIENCE

Each part of the eye has a different job. All of these jobs involve focusing light energy to create visible images.

The Cornea

Our eyes are covered with a clear tissue called the *cornea*. It bends the light as it enters our eyes.

The Iris

A ring-shaped muscle called the *iris* is located near the front of the eye. This muscle can be many different colors. If you have blue eyes, your iris is blue. Brown-eyed people have brown irises. Like other muscles in your body, the iris can **expand** and **contract**. This causes the opening inside the ring to grow larger or smaller.

The Pupil

The opening in the center of the iris is called the *pupil*. The pupil's job is to control the amount of light that your eyes let in. The muscles in the iris change the size of the pupil to let in the proper amount of light.

When the pupils are open wide, lots of light can enter your eyes. When you enter a dim room, the iris muscles expand, making your pupils larger. When you are outside and it's getting dark, your pupils get larger. Your eyes need more light to see in these situations, so the iris opens the pupils to let in as much light as possible.

When the pupils are small, not much light can enter your eyes. When a bright light shines in your eyes, the iris muscles contract, making the pupils smaller. When you are outside on a sunny day, your pupils get smaller. Your eyes allow just enough light in for you to see. Light that is too bright could hurt your eyes.

The Lens

Light passes through the lens of each eye next. The lens in each eye is convex in shape. So when light passes through the lens, it bends inward. The light then travels through the watery fluid in your eyes. The light rays cross over and fall back on the far surface of the eyeball.

Reflective Retinas

The retina in some animals is very reflective, like a shiny mirror. This is why you can sometimes see a cat's or dog's eyes "glow" in the dark. The light is bouncing off its retinas and reflecting back at you.

The Retina

The back surface of the eye is called the *retina*. When light strikes this surface, an image of the object you're looking at appears. The image is inverted, or upside down, because the light rays crossed in their travel.

The retina is filled with cells called *rods* and *cones*. Rods sense light and dark. Cones sense colors. These cells help us see shades and different colors.

The Brain

An **optic nerve** connects each eyeball to the brain. The optic nerves send the visual information to the brain. The brain then makes sense of what the eyes are seeing. For example, the brain flips over the image formed on the retina. Now it appears right side up.

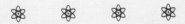

Working together with your eyes, light energy creates all of the images that you see!

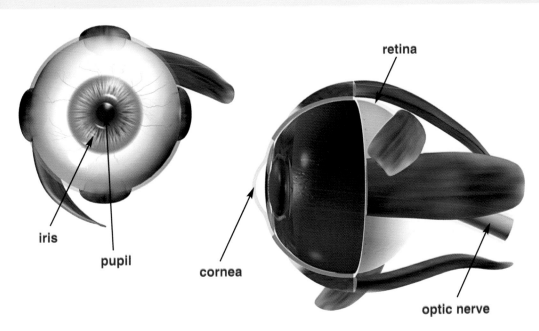

iris

pupil

cornea

retina

optic nerve

Lenses Correct Light Problems

People whose eye lenses bend the light rays so they fall in front of the retina are called *nearsighted*. They can see objects that are near, but images that are farther away are blurry. Concave lenses help the light rays reach the retina.

Farsighted people have lenses that bend the light so it falls behind the retina. They can see objects that are far away, but up close images are blurry. Convex lenses help the light rays come together at the retina.

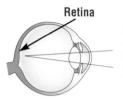
Retina

Nearsightedness occurs when light rays meet before they reach the retina.

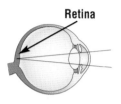
Retina

Farsightedness occurs when light rays meet beyond the retina.

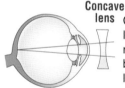
Concave lens

Concave lenses correct nearsightedness by bringing the light rays together at the retina.

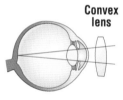
Convex lens

Convex lenses correct farsightedness.

Try This!

Borrow a pair of eyeglasses from a classmate or teacher. Use an overhead projector and screen. Turn on the projector and hold the glasses between the screen and the light. Do the glasses focus the light into a bright spot in the center? Or do they spread out the light into a bright ring, leaving a dark area in the center?

Determine whether the lenses are concave or convex. If the lenses bring the light together, they are convex. If the light spreads, the lenses are concave.

Is the owner of the glasses farsighted or nearsighted?

CHAPTER 9
Colors

A RAINBOW OF COLORS

The light from the Sun and from lamps is called *white light*. This white light is really made up of many different colors.

In the 1600s, Isaac Newton proved this theory by using a glass prism to separate white light into the colors of the rainbow. A prism is a piece of glass or plastic that separates light into its colors. These colors are red, orange, yellow, green, blue, indigo, and violet. Newton then used a second prism to combine the colors back into a single beam of white light.

A prism can separate colors because each color has a different wavelength. Red light has the longest wavelength. Orange is a bit shorter. From yellow to green to blue light, the wavelengths continue to decrease in size. Violet has the shortest wavelength that our eyes can see. This range of wavelengths and colors is part of the *electromagnetic spectrum*.

Indigo Information

Indigo is a deep reddish blue color that falls between the blue and violet colors in a rainbow.

When white light enters a prism, it bends due to refraction. The glass or plastic of the prism is denser than air. The light slows down and bends. Because each color making up the white light is a different wavelength, each color is bent differently. Red bends the least. Violet bends the most.

Sometimes sunlight and rain create a natural rainbow. The little droplets of rainwater act as tiny prisms for sunlight shining at a low angle. The light enters the droplets and slows down. Each color in the sunlight bends and separates, producing a rainbow in the sky.

Meet ROY G. BIV

An easy way to remember the colors of the rainbow in the order they appear is to remember the name ROY G. BIV.

- <u>R</u>ed • <u>O</u>range • <u>Y</u>ellow
- <u>G</u>reen
- <u>B</u>lue • <u>I</u>ndigo • <u>V</u>iolet

PIGMENTS AND DYES

Pigments and dyes give color to materials. Pigments are an object's own natural colorings. Dyes are artificial colors that can be added to fabrics, plastics, paints, and food.

Make your own rainbow. Turn on a hose on a sunny day. Adjust the water spray to a fine mist. Can you see a rainbow?

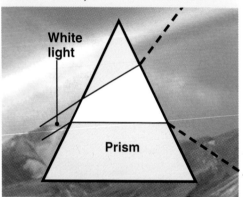

White light

Prism

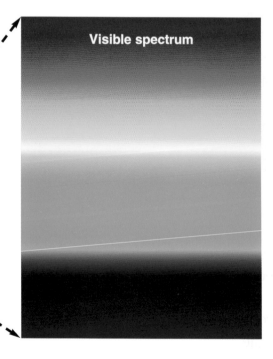

Visible spectrum

We see objects as different colors depending on the pigments or dyes in those objects. Some of the light that hits the object is absorbed. Some of the light is reflected. The pigments or dyes used to color the object absorb all the colors that make up light except the color(s) of the pigment or dye. That color is reflected back to our eyes.

Plants have green pigments. When light strikes a plant, the plant absorbs all of the colors except green. The green light is reflected back to our eyes, so the plant appears green.

A red fire engine absorbs all the colors of light except red. The red pigments used in the paint reflect the color red back to our eyes.

The dye used in making blue jeans absorbs all the colors of light except blue. The blue color is reflected, giving the jeans a blue color.

When all the colors in light reflect back to our eyes, we see an object as white. Objects appear black when none of the colors reflect from the item. Black objects absorb all of the color waves.

INFRARED AND ULTRAVIOLET LIGHT

Other forms of light exist that are not seen as colors. These types of light are called *infrared* (below red) and *ultraviolet* (above violet). Infrared light has a longer wavelength than the color red. Ultraviolet, or UV, waves are shorter than violet ones.

Human eyes cannot see these two forms of light. Special tools help us detect them.

Night vision goggles allow us to see infrared light. The goggles pick up heat given off by objects. Hotter objects show up brighter than cooler objects.

Firefighters can see through thick black smoke with infrared cameras. They can find people trapped inside burning buildings much faster with this tool.

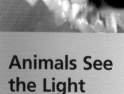

The Earth's ozone layer protects humans from most of the ultraviolet light given off by the Sun. To study this light, scientists send **satellites** into outer space. Telescopes that can "see" the UV light collect information and take pictures of stars and **galaxies**.

A special UV light called a *black light* can be used to detect objects with UV rays. This black light gives off a purple light that humans can see. It also gives off invisible UV rays. Materials containing chemicals called *phosphors* glow under the black light. The UV rays give the phosphors energy that they turn into light we can see.

Many detergents have phosphors in them that make white clothing glow under a black light. Black lights can be used to find chemical leaks in air conditioners and refrigerators. Ultraviolet lights are also used to help solve crimes. Fingerprints and bloodstains show up under a UV light.

Animals See the Light

Some animals can see infrared light, while others can see ultraviolet light. Most insects can see ultraviolet light.

Rattlesnakes can spot their prey because of the infrared energy (body heat) the prey give off. Goldfish can see in the dark because of their infrared vision. Bees use their ability to see ultraviolet light to find flowers, which provide food energy for the insects.

Remember Your Sunglasses and Sunscreen!

While humans cannot see ultraviolet light, they can definitely see and feel its effects. UV rays from the Sun are what cause tans and sunburns. Too much exposure to UV light without protection can cause skin cancer in humans.

CHAPTER 10
Tools of Light

Light energy is important to many tools in today's world. Some of these tools provide light. Others use light energy to perform a variety of jobs.

LASERS

Lasers use concentrated light of only one wavelength and color. Red is the most common laser color, but there are other colored laser lights as well.

When all the light present is the same wavelength and color, the light becomes very powerful. Because of this, lasers can be used as cutting tools. Some surgeries are done using lasers. The very fine beam of light can cut very accurately and precisely.

Some lasers release an extremely powerful infrared (heat) beam. These lasers are used on fabrics, metals, and rock. The

beams can cut patterns out of hundreds of layers of fabric at a time. This makes the process of making clothes faster, easier, and more cost efficient. Big industries also use lasers for cutting, drilling, and welding.

Lasers are used in equipment around homes and schools. Compact disc players and DVD players use a laser to read information. The laser light reads sound and visual information from the discs.

Stores also use laser light. The price scanner at the checkout uses a laser to read the bar codes on products.

CAMERAS

The camera is another invention that uses light. A camera works much like your eyes. An opening called the *shutter* lets light into the camera. This is similar to the job your pupils perform. A lens in the camera bends the light and focuses the image on the film. These tasks are like those performed by your eye lenses. The film is sensitive to light just as your retinas are.

To record an image, the film undergoes a chemical reaction with the light. The film undergoes another chemical reaction when it is developed into pictures.

MICROSCOPES

Light microscopes were invented about 400 years ago. These tools use light and lenses to help us see small objects better. The lenses bend the light rays shining through the object. These rays then join and magnify, or enlarge, the image so it appears bigger. This allows us to spot details on small objects that are not detectable to the human eye.

The size that the object appears depends on the power of the lenses used. A compound microscope has two lenses. The ocular, or eyepiece, lens magnifies the object to 10 times its size.

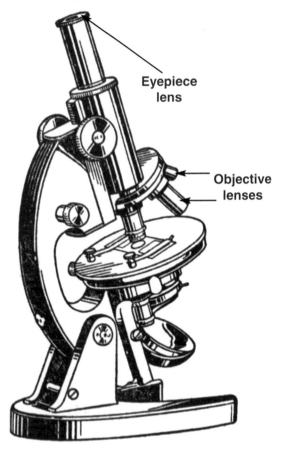

Eyepiece lens

Objective lenses

TELESCOPES

Telescopes were also invented about 400 years ago. Telescopes help us see objects at a great distance by collecting light. Reflecting telescopes use mirrors, and refracting telescopes use lenses. These large mirrors or lenses gather large amounts of light in long tubes. The eyepiece lens then helps focus the distant object into view.

Out of This World

The Hubble Space Telescope has been located in outer space since 1990. It uses a huge mirror to focus on objects. Scientists receive pictures and other information collected by the telescope.

The **objective** lens can magnify the object anywhere from 4 to 100 times. By multiplying the **magnification** of the eyepiece lens times the objective lens, you get the total magnification. A typical compound microscope can magnify an object 40 to 1000 times the actual size of the object.

Doctors, lab technicians, and scientists often use microscopes that are much more powerful. These advanced microscopes can magnify objects to thousands of times their actual size.

Telescopes allow people to see close-up views of planets, moons, and other **celestial** bodies. Just as with microscopes, the power of telescopes varies greatly. Scientists have much stronger telescopes than those used in homes for stargazing.

FIBER OPTICS

Light can also help send signals across distances. Optical fibers are thin, flexible wires made of pure glass or plastic. These wires contain a mirrorlike surface. Light can be sent in signals starting at one end of the wire. The light bounces back and forth inside the glass wire. It quickly reaches the other end of the wire. At this end, the light signal can be changed into electrical signals that make sense to computers, televisions, and telephones.

Fiber optics are also important in several professional fields. Optical fibers attached to cameras can transfer images from one end of a group of wires to the other. These cameras are used by doctors to explore body organs. Engineers use fiber optics to inspect pipes and engines. Plumbers use these special wires to check out sewer lines.

Many fibers bundled together can carry large amounts of sound and visual information. Because of this, the use of fiber optics is cheaper and more efficient than copper wires for carrying information over long distances.

❋   ❋   ❋   ❋

Thousands of scientists and inventors have experimented with both natural and artificial light. Their work has left us with a better understanding of how light works. It has also paved the way for the many tools and machines that use light to improve our lives. Our lives are definitely brighter thanks to light energy!

Fiber optic cable

Light Signal 1 ——

Light Signal 2 ——

Mirrorlike wall

The light in a fiber optic cable travels by bouncing back and forth off the mirrorlike wall of the cable. This is called *total internal reflection*.

INTERNET CONNECTIONS AND RELATED READING FOR LIGHT ENERGY

Energy Quest
(http://www.energyquest.ca.gov/index.html)
This fun site provides information, stories, news, projects, games, and links to other energy sites. Visit the Gallery of Energy Pioneers too.

How Stuff Works
(http://www.howstuffworks.com)
If you have questions about how anything involving energy (and anything else!) works, this Web site is the place to look. It includes sections on energy and electrical power, light, and many other inventions related to energy.

Energy Information Administration
(http://www.eia.doe.gov/kids/index.html)
Review the definition of energy and its forms here. Then check out the Kid's Corner, Fun Facts, and Energy Quiz.

U.S. Department of Energy
(http://www.eren.doe.gov/kids/)
Dr. E's Energy Lab will teach you about solar energy and energy efficiency. A dog named Roofus shows you his energy-efficient home and neighborhood. Many links to other energy sites can be found here.

The Atoms Family
(**http://www.miamisci.org/af/sln**)
This spooky Web site teaches about different forms of energy through simple experiments.

One World.Net's Kid's Channel
(**http://www.oneworld.net/penguin/energy/energy.html**)
Tiki the Penguin discusses the positive and negative sides of different types of energy sources.

Day Light, Night Light: Where Light Comes From by Franklyn M. Branley. Discusses the properties of light, particularly its source in heat. HarperCollins, 1998. [RL 2.3 IL 2–4] (5666601 PB 5666602 HB)

Energy by Jack Challoner. An Eyewitness Science book on energy. Dorling Kindersley, 1993. [RL 7.9 IL 3–8] (5868606 HB)

Energy by Alvin and Virginia Silverstein and Laura Silverstein Nunn. Explains a fundamental concept of science, gives some background, and discusses current applications and developments. Millbrook Press, 1998. [RL 5 IL 5–8] (3111906 HB)

Light by David Burnie. An Eyewitness Science book on light. Dorling Kindersley, 1992. [RL 8.5 IL 3–8] (5869106 HB)

Light by Darlene Lauw and Lim Cheng Puay. Presents activities that demonstrate how light works in our everyday lives. History boxes feature the scientists who made significant discoveries in the field of light. Crabtree Publishing, 2002. [RL 3.8 IL 2–5] (3396701 PB)

- RL = Reading Level
- IL = Interest Level

Perfection Learning's catalog numbers are included for your ordering convenience. PB indicates paperback. HB indicates hardback.

GLOSSARY

absorb (uhb SORB) take in

atom (AT uhm) tiny particle that makes up all materials in the universe (see separate entry for *particle*)

axis (AK SIS) imaginary line that runs through the Earth from the North Pole to the South Pole

celestial (suh LES chuhl) relating to the sky, outer space, or the heavens

chemical (KEM uh kuhl) substance produced by a scientific reaction; substance used to create a scientific reaction; relating to either of these substances

contract (kuhn TRAKT) grow smaller; close up

current (KER uhnt) flow of electrical charge

expand (ik SPAND) grow larger; open wider

galaxy (GAL ek see) large group of stars

hemisphere (HEM uh sfear) half of the Earth

infrared (in fruh RED) light outside of the visible spectrum on the red side

irregular reflection (ir REG yuh ler ree FLEK shuhn) light bouncing off a rough, uneven, bumpy surface

magnification (mag nuh fi KAY shuhn) measurement of the enlargement of an object

mate (mayt) reproduce; create babies

moderate (MAH duh ruht) not extreme; falling in the middle

objective (ob JEK tiv) relating to a lens or system of lenses that forms an image of an object

optic nerve (OP tik nerv) nerve that connects the eye to the brain

particle (PAR tuh kuhl) small piece or amount of something

regular reflection (REG yuh ler ree FLEK shuhn) light bouncing off a flat, smooth, shiny surface

revolve (ree VOLV) to move in a curved path around a center or axis (see separate entry for *axis*)

rotation (roh TAY shuhn) turning around a center or axis (see separate entry for *axis*)

satellite (SAT uh leyet) human-made object or vehicle that orbits in space

shade (shayd) lighter or darker variation of a pure color of the spectrum

theory (THEAR ee) set of beliefs or facts usually proven or disproven by scientific experiment

transparent (tranz PAIR ent) allowing light to pass through; able to be seen through

turbine (ter bin) engine that turns from the power of water, steam, light, or air

ultraviolet (uhl truh VEYE let) light outside the visible spectrum on the violet side

wave (wayv) movement of energy in a continual up-and-down pattern

INDEX

Legends of the Sea

Sea Monsters

Catherine Veitch

Chicago, Illinois

www.heinemannraintree.com
Visit our website to find out more information about Heinemann-Raintree books.

To order:
☎ Phone 888-454-2279
💻 Visit www.heinemannraintree.com to browse our catalog and order online.

©2010 Raintree
an imprint of Capstone Global Library, LLC
Chicago, Illinois

Edited by Rebecca Rissman, Nancy Dickmann, and Siân Smith
Designed by Joanna Hinton Malivoire and Ryan Frieson
Original illustrations ©Capstone Global Library 2010
Original illustration p.29 ©Steve Walker
Illustrated by Mendola Ltd and Steve Walker
Picture research by Tracy Cummins
Production control by Victoria Fitzgerald
Originated by Capstone Global Library Ltd
Printed and bound in China by Leo Paper Products Ltd

14 13 12 11 10
10 9 8 7 6 5 4 3 2 1

Library of Congress Cataloging-in-Publication Data
Veitch, Catherine.
Sea monsters / Catherine Veitch. -- 1st ed.
p. cm. -- (Legends of the sea)
Includes bibliographical references and index.
ISBN 978-1-4109-3788-9 (hc)
ISBN 978-1-4109-3793-3 (pb)
1. Sea monsters--Juvenile literature. I. Title.
QL89.2.S4V45 2011
591.77--dc22
                    2009045674

Acknowledgments
The author and publishers are grateful to the following for permission to reproduce copyright material: akg-images pp.9 (© Peter Connolly), 25; Alamy p.6 (© Adam Silver); AP Photo p.19 (NZPA/Ross Setford); Art Resource, NY p.7 (Erich Lessing); Corbis p.23 (© Denis Scott); Getty Images p.18 (Ministry of Fisheries); istockphoto pp.26 left (© Philip Roop) 26 right (© ido hirshberg), 27 (© natsmith1); National Geographic Stock pp.11, 20 (Minden Pictures/ Norbert Wu), 13, 14 (Paul Sutherland); Photolibrary p.12 (Karen Gowlett-Holmes); Shutterstock p.15 (© Niar), 21 (© ampower); The Bridgeman Art Library International p.17 (Bibliotheque des Arts Decoratifs, Paris, France / Archives Charmet).

We would like to thank Steve Walker for his invaluable help in the preparation of this book.

Some words are shown in bold, **like this**. You can find out what they mean by looking in the glossary.

Contents

Is It True?

Legends, or stories, have been told about strange creatures that live in the sea. Do you know which of these monsters are real and which are not?

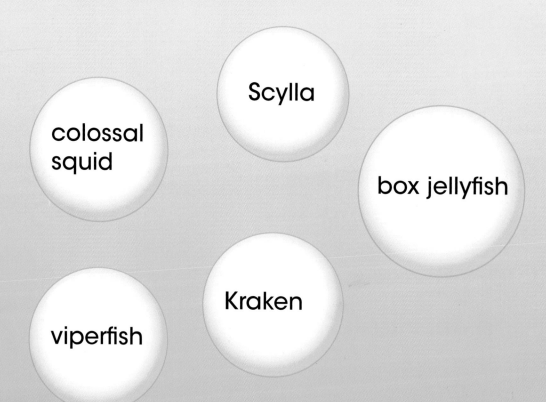

colossal squid

Scylla

box jellyfish

Kraken

viperfish

Read on to meet these sea monsters and others. Find out if you are right.

These signs will tell you if the sea monsters are real, not real, or if we don't know.

real

not real

don't know

In a Spin

Long ago, people in Greece told stories of a sea monster called Charybdis [say car-ib-dis]. They say she created giant **whirlpools** by swallowing and burping up seawater.

In a whirlpool, water swirls in a circle and can pull things under.

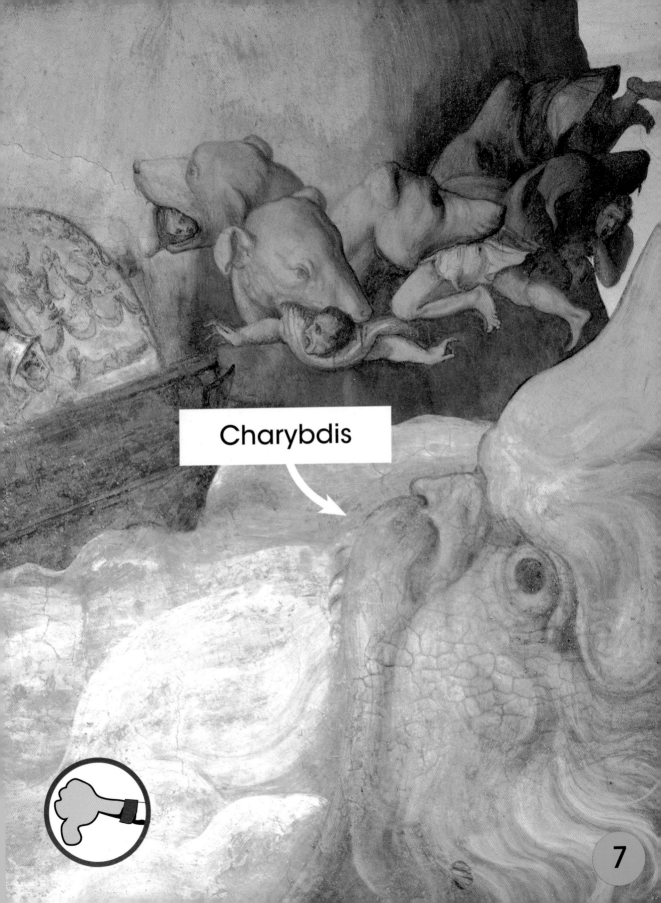

Charybdis

The monstrous sea goddess Scylla [say sill-uh] lived near Charybdis. Scylla had twelve dog's legs and six huge dog's heads. She had three rows of teeth in each head !

Some **legends** say that Scylla actually had six heads on snake-like necks.

9

Fierce Fangs

The viperfish is only about 1 foot long. That's about as long as two pens put end to end. But it has fierce **fangs**, or teeth, that make this small fish extra scary.

fangs

The viperfish swims toward its **prey** at high speed and then attacks with its fangs.

prey

viperfish

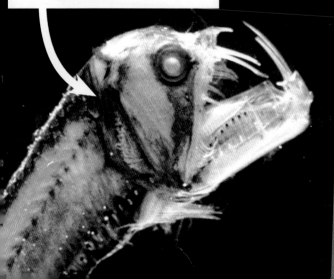

IS IT TRUE?

The viperfish gets its name because it has a body and fangs like a viper snake.

Answer: true

11

Super Stingers

Some types of box jellyfish are among the most **poisonous** creatures on Earth. Their name comes from their box shape. They have about 15 **tentacles** on each corner and each tentacle can have thousands of stinging cells!

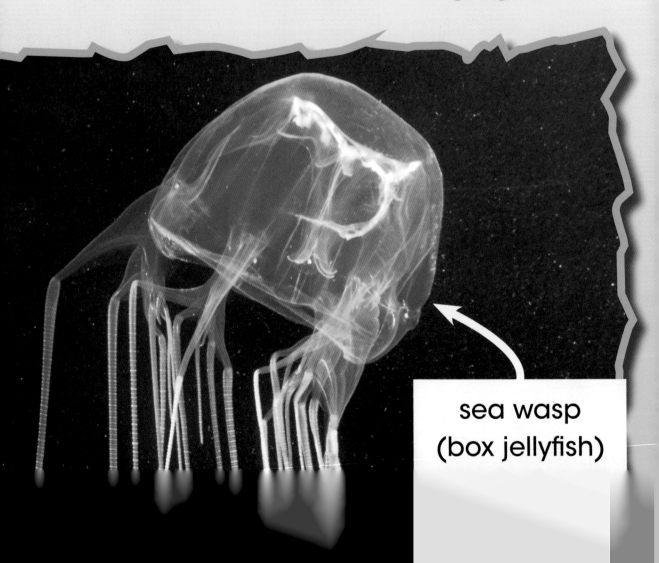

sea wasp
(box jellyfish)

tentacle

Once stung by a sea wasp box jellyfish, you could have only minutes to live. Their stings are incredibly painful. **Victims** may die of a heart attack before even reaching land. People need to get help quickly to survive this scary attack!

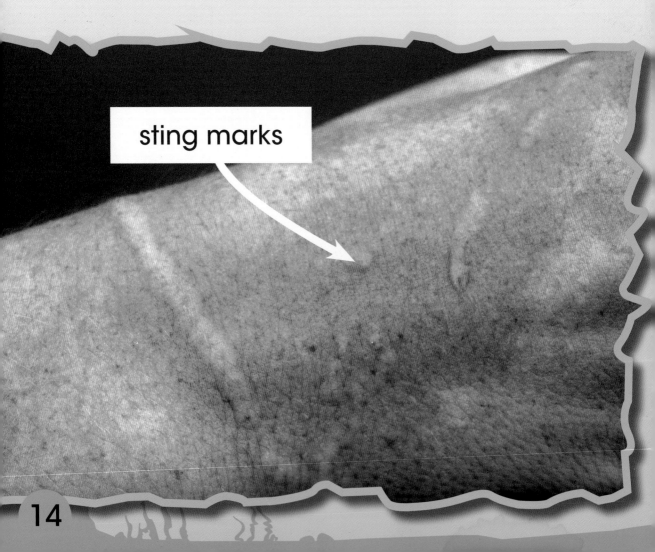

sting marks

vinegar

MARINE STINGERS
ARE PRESENT
IN THESE WATERS
DURING THE
SUMMER MONTHS

VINEGAR

MARINE STINGS
DO NOT RUB
CAL ATTENTION

DID YOU KNOW?
Pouring vinegar over a
box jellyfish sting can
help to stop the **poison**.

15

Sirens' Song

Some ancient Greek **legends** told the story of the Sirens. When sailors heard the Sirens' beautiful singing they could not turn away. They steered their ships toward the Sirens. The ships were smashed to pieces on rocks.

The best way to survive a Siren's singing is to cover your ears!

Siren

Odysseus

DID YOU KNOW?
Legends say Odysseus heard the Sirens' song and lived. His men tied him up so that he couldn't steer the boat. Then they blocked up their ears.

Shy Squid

A few years ago a colossal squid was caught near New Zealand. This giant squid weighed 1,091 pounds. That's about as heavy as seven men! Only a few colossal squid have ever been found. None of these have been alive.

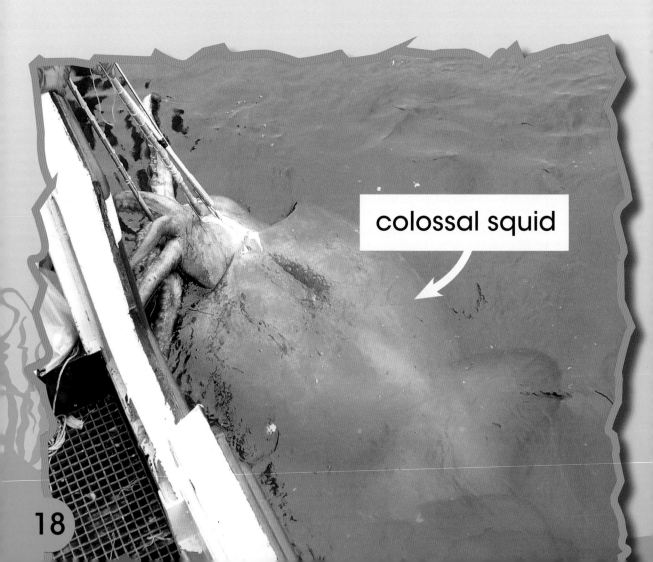

colossal squid

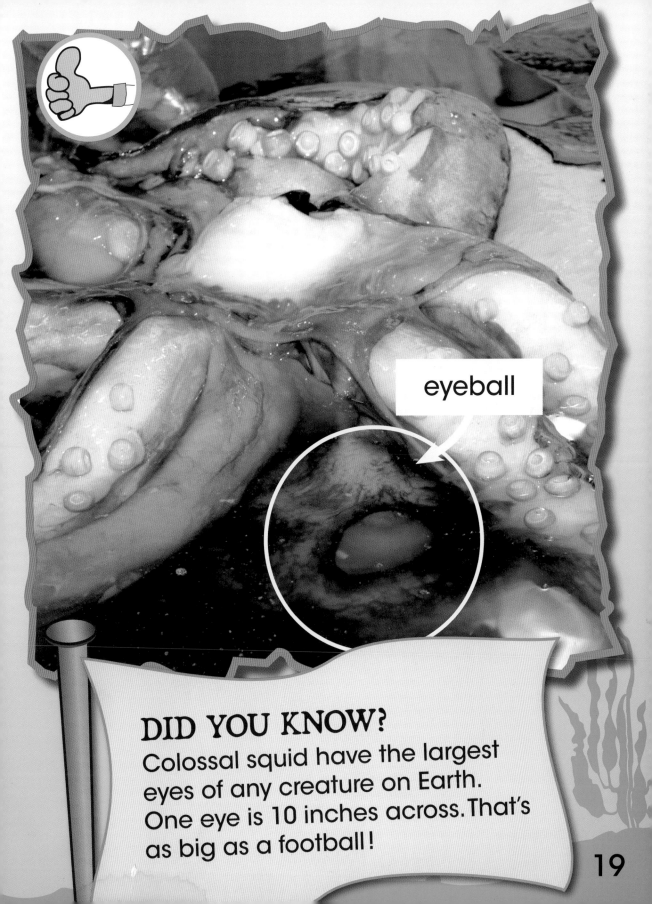

eyeball

DID YOU KNOW?

Colossal squid have the largest eyes of any creature on Earth. One eye is 10 inches across. That's as big as a football!

Greedy Gulper

The weird looking gulper eel has a huge mouth. It can open its mouth wide enough to swallow **prey**, or creatures, much larger than itself. Its stomach can also stretch to fit in a bigger meal!

The gulper eel has a light at the end of its whip-like tail, which attracts prey toward its mouth.

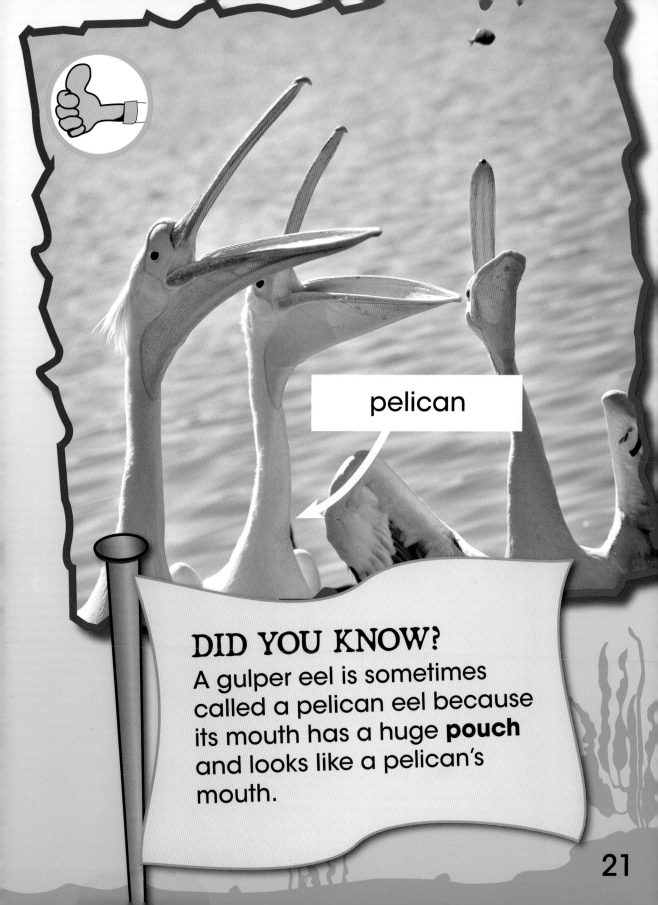

pelican

DID YOU KNOW?
A gulper eel is sometimes called a pelican eel because its mouth has a huge **pouch** and looks like a pelican's mouth.

Noisy Neighbor

The blue whale is the loudest animal on Earth. It is even louder than an airplane. Its whistle can be heard under the ocean for hundreds of miles.

DID YOU KNOW?

The blue whale is the largest animal that ever lived. It grows to about 80 feet long. That's as long as two buses put end to end!

Crazy Kraken

Legends of the Kraken describe a sea monster that is so big that it is mistaken for an island. Some stories say that sailors who land on the Kraken are sucked underwater. Others say that the Kraken's huge **tentacles** drag whole ships underwater.

tentacle

DID YOU KNOW?
Some people think the Kraken is actually a colossal squid and that some of the stories about it may be true.

Monster Hunters

People who study and search for **legendary** creatures are called **cryptozoologists** [say crip-toe-zoo-ol-o-gists]. Who knows what monsters lurk in the deep, dark depths of the world's oceans?

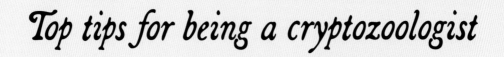

Top tips for being a cryptozoologist

1. Read up on your **legends**.
2. Make a list of any legendary creatures that you think could be real.
3. Learn how to **track** and spot animals, and how to collect **evidence**.
4. Be patient. Remember that at first, no one believed that colossal squids were real.

Imagine you are a **cryptozoologist** and come face-to-face with a sea monster. Draw what you imagine the monster would look like.

Make a table like this to help you plan your monster first. Use your imagination to fill out the columns.

Special features	Lives	Eats
large eyes, sharp teeth, three tentacles	under rocks	people

Give your monster a name.

Croctopus

Glossary

cryptozoologist person who investigates legendary or mystery creatures to find out if they are real or not

evidence information used to prove something

fangs sharp, pointed teeth

legend story that started long ago. Legends can be true or made up.

legendary something that comes from a legend

poisonous dangerous to humans or other animals. Poison can cause illness or death.

pouch part of an animal's body where things can be kept. Similar to a pocket.

prey animals that are hunted and killed for food

tentacle long, thin part of a creature's body similar to an arm without a hand on the end

track follow

victim person or group of people who are harmed

whirlpool place in the sea or a river where water swirls around in a circle. Sometimes whirlpools are strong enough to drag things under the water.

Find Out More

Books

Pipe, Jim. *Scary Creatures of the Deep*, Franklin Watts, 2009.

Ryan-Herndon, Lisa. *Planet Earth: Deep Ocean Creatures*, Scholastic, 2009.

Woodward, John. *Discoverology Series: Creatures of the Deep*, Barron's Educational Series, 2009.

Websites

atschool.eduweb.co.uk/carolrb/greek/greek1.html
Learn about ancient Greek sea monsters.

news.bbc.co.uk/cbbcnews/hi/pictures/galleries/newsid_3881000/3881353.stm
Find out interesting facts about jellyfish on this website.

www.enchantedlearning.com/subjects/whales/
This website has many facts on the blue whale.

www.seasky.org/deep-sea/deep-sea-menu.html
Take on the role of a deep sea explorer on this website. Click on the creatures you discover to find out more about them.

Find out

Which creature has no heart?

Index